CW00502467

Fire Engines – a Colour Portfolio

Roger C. Mardon

Right: Leyland Cub pump escape new to Yeovil, Somerset in 1936.

Picture Credits

All photographs are by the author, who has a long-standing interest in fire brigade vehicles which he has been photographing for many years. He is a lifelong resident of Kent and is married with two grown-up sons.

First published 1999

ISBN 0 7110 2639 4

Published by Ian Allan Publishing
an imprint of Ian Allan Publishing Ltd, Terminal House, Shepperton, Surrey TW17 8AS.
Printed by Ian Allan Printing Ltd, Riverdene Business Park, Hersham, Surrey KT12 4RG.

Code: 9904/B2

Introduction

Ever since Prometheus stole fire from the Gods, man has been devising ways of controlling it. Under control, fire has always been of great domestic and industrial benefit. But out of control it is one of the most destructive forces known to the world.

Whilst buckets of water were the primary weapons in the fight against fire unleashed, the Romans used force pumps for fire-fighting over 2,000 years ago. Nothing better was developed for centuries, and syringes and squirts were still at the leading edge of technology until the 17th century. After an absence of over 1,000 years, pumps capable of producing a constant, rather than intermittent, stream of water were redeveloped and versions mounted on wheels were built. These were powered by men working levers or treadles and could be fed with water either from their own cistern or by suction from an open water source. Larger horse-drawn manual pumps for working often by 20 men, and sometimes as many as 50, were designed in the 18th and 19th centuries.

It was not until 1829 that the first steam pump was built. This, too, was horse-drawn and, after initial resistance, steam power predominated until the turn of the century and beyond. Self-propelled steamers were developed but were rapidly superseded by the introduction of the petrol engine. The first motor fire engines were little more than large chemical fire extinguishers on wheels but in 1904 the English firm of Merryweather & Sons built the first appliance where the petrol engine powered both the vehicle and its pump. In the same year Manchester took delivery of the country's first turntable ladder, a horsed carriage upon which was mounted a German-designed 82ft (25m) extending ladder capable of being rotated through 360° and powered by compressed gas. Two years later the first motorised turntable ladder had been built.

Ladders have long been associated with fire-fighting and rescue from fire. They were sited at strategic points in cities and towns to be carried to the fire and, in the first half of the 19th century, escape ladders were mounted on a two-wheeled carriage for greater manoeuvrability. It was not unusual for horse-drawn manual and steam pumps to carry a simple wooden ladder and horse-drawn vehicles were developed to carry the wheeled escape. By the beginning of World War 1 motor pumps carrying extending ladders or wheeled escapes had become established, although by no means universal. There, probably, we have the beginning of the fire engine as we know it today.

The motor vehicle evolved between the two world wars and the fire engine, or appliance as it is properly called, evolved with it. Designs continued to reflect the horse-drawn era with firemen precariously sitting or standing along the sides of 'Braidwood' style open-bodied pumps, so called after James Braidwood, first Superintendent of the London Fire Engine Establishment between 1833 and 1861. Some appliances were arranged where the crew faced forward on seats across the vehicle, 'transverse' style, and some where firemen sat down the sides on inward facing seats, 'New World' style. Pumping appliances carrying a ladder typically extending to 30ft (9m) were called pumps and those carrying a

Front cover: The Leyland FK9 with midships-mounted pump, and the FK10 with rear-mounted pump, were in production from 1938 to 1940. This FK9 pump escape with limousine bodywork was built in 1939 for Bournemouth Fire Brigade. It has a 700gpm pump and is fitted with an 80gal water tank. Bournemouth, a county borough from 1900, was geographically within Hampshire until the local government reorganisation of 1974 which redrew the county boundary. This south coast resort then became a district of Dorset and ceased to be responsible for its own fire brigade.

Back cover: The Auxiliary Fire Service (AFS) was equipped not only with pumps but also with a range of special and ancillary appliances, including pipe carriers which carried ⅓mile of 6in piping to maintain water supplies, hose layers and foam tenders. Unique to the AFS was the transportable water unit, or 'Bikini' unit, designed to pump water from sources not readily accessible to land appliances. Each vehicle carried three inflatable rafts and nine light portable pumps. The rafts, each with three pumps, could be launched across soft ground or lowered with the vehicle's crane into the water from where they could pump water ashore for fire-fighting purposes. This unit was built by Mann Egerton in 1954 on a Commer Q4 four-wheel-drive chassis.

wheeled escape, usually 50ft (15m), were called pump escapes. By the early 1930s the enclosed limousine pump had appeared, offering some concession to safety and comfort, although some were without glass in the side windows. Turntable ladders were usually capable of extending to 100ft (30m), sometimes 125ft (38m) and occasionally 150ft (46m).

Pumping appliances carried hose, nozzles and other gear for fire-fighting purposes but equipment not regularly used would often be carried on special vehicles known as emergency tenders. Frequently, these were developed to meet local needs but would be likely to provide a generator and searchlights, rescue gear such as cutting equipment and hydraulic jacks, extra breathing apparatus sets, and even a smoke exhauster to ventilate smoke-logged premises. Foam tenders, to tackle oil fires, had been in use since 1920 and salvage tenders, equipped to protect property from water damage and the effects of fire-fighting as much as to rescue it from the fire, even longer. Some fire brigades operated hose layers capable of paying out two lines of hose, perhaps a mile long, while travelling at about 20mph.

World War 2 created a need to produce fire pumps in great numbers and utility appliances on various commercial chassis were built. Trailer pumps were built in huge numbers to be towed by

Below left: London manufacturer Richard Newsham first patented his fire engine in 1721 and this model dates from about 1730. The hardwood machine essentially comprises a wheeled tank into which water was emptied from buckets. Two single-acting pump barrels in the tank were brought into action by one or two men on each side working the levers up and down. A stream of water pumped from the tank, or drawn by suction from an open supply, was discharged from the pipe on top of the engine which could be swivelled in the direction of the fire. Provision of an air vessel ensured a continuous flow of water which could also be delivered through a leather hose.

Below: The maker of this hand-drawn manual pump is unknown but it was purchased in 1755 at a cost of £13 4s 9d for Martock in Somerset, where it was kept in the porch of All Saints Parish Church. The twin reciprocating pump, with 3in bore and 8in stroke, was capable of producing 33gpm and could deliver a jet of water to a height of 70ft. This engine was last known to be in use at a thatch fire at Spillers Bakery, High Street, Martock in 1899 when it was carried through the back door and pumped from an indoor well.

Left: Shand Mason & Co Ltd of Blackfriars, London, founded in 1851, was one of Britain's best known fire engine builders with origins going back to 1774. This hand-drawn hose-reel carrier was constructed by the company in 1880 and comprises a wooden hose-reel mounted between the two wheels of a carriage. The box on top would have been used to carry equipment such as branches, nozzles and standpipe. The example shown was operated by the Norwich Police Fire Brigade.

Right: Faversham was served by two insurance brigades and a volunteer 'gentlemen's brigade' before the corporation formed its own fire brigade in 1926. This 1880 Shand Mason 22-man manual pump, which would have been operated by 11 men working the levers on each side, originally served with the Norwich Union brigade before being taken over by the town council. It was last used for fire-fighting at a skating rink fire in 1926 and soon after was replaced by a motor pump. In World War 2 it was used as an obstacle on open ground to block enemy aircraft and parachute landings. It was restored in 1969.

light vehicles. Frequently taxis were used before purpose-built auxiliary towing vehicles became widely available. At the beginning of the war London alone had some 2,000 extra pumps available to the Auxiliary Fire Service. Hostilities with Germany brought to an end the import of Magirus and Metz turntable ladders, leaving the construction of new appliances and the development of the wartime 60ft (18.25m) hand-operated turntable ladder to Merryweather.

Utility versions of other prewar regular appliances were built but wartime fire-fighting generated needs not known in peacetime and special vehicles were provided to meet those demands. Worthy of special mention is the mobile dam unit. A dam is a water reservoir, and steel-framed collapsible dams, self-supporting canvas dams or galvanised iron tanks of 500gal (2,273 litres) or more were mounted on flat-platform lorries, which then became known as dam lorries. Such a lorry towing a trailer pump or with a pump mounted on the platform was known as a mobile dam unit, and this was the forerunner of the water tender, the mainstay of today's fire service.

In the early years after the war, many fire engines were adapted or rebodied for peacetime use and, along with prewar regular appliances, these continued to give service until the 1960s. New designs emerged in the late 1940s and within a few years brigades up and down the country were renewing their fleets. In urban areas the pump and pump escape continued to prevail but brigades with more rural risks started using water tenders with their increased water carrying capacity. These were often built on standard or modified commercial chassis by manufacturers specialising in fire engine bodies and equipment, but fire engines purpose-built from the chassis up, by the likes of Dennis Bros, remained highly regarded and continued in production. A postwar innovation was the 45ft (13.5m) ladder which gradually superseded the wheeled escape and gave rise to a new fire appliance designation. The principal ladder carried on a pump or water tender was 30ft or 35ft (9m or 10.5m) and so, to distinguish them, appliances with a 45ft ladder were designated pump ladders or water tender ladders. The wheeled escape has now completely disappeared from service and in some brigades a 45ft ladder (commonly known as a 1-3-5, from its metric size) is now found on all pumping appliances.

With the advance of technology, turntable ladders, along with all fire engines, became more sophisticated but the biggest change in the design of aerial appliances came in the early 1960s with the introduction of the hydraulic platform. This takes the form of a cage elevated on two or three booms which, because they are articulated, can reach up over obstructions and down the other side. In some brigades hydraulic platforms superseded the turntable ladder but in many the advantages of each are called for and the two types work alongside each other. A more recent development is the aerial

ladder platform which combines advantages of both, with a ladder running alongside telescopic articulated booms. Even an articulated turntable ladder has now appeared.

As evidenced by the changing nomenclature from fire brigade to fire and rescue service, a high proportion of a fire-fighter's work is the rescue of people from dangerous situations not involving fire and the extrication of people trapped in road accidents. It was found that large emergency tenders were not always suited to this work so lighter and faster rescue tenders were developed. The trend now is for rescue equipment to be carried on pumping appliances, giving rise to the inclusion of the word 'rescue' in appliance designations, eg rescue pump. Industrial advances have brought greater risks and the widespread use and transport of chemicals and other hazardous materials spawned a new type of fire appliance — the chemical incident unit. Such units provide for the protection and decontamination of fire-fighters involved in an incident and some also carry equipment necessary to neutralise and remove dangerous materials. In some brigades heavy rescue and other specialist gear is still carried on emergency tenders, now more often called rescue tenders, and elsewhere various combinations of this equipment are carried on vehicles known as incident or operational support units.

Road tankers have been adapted or specially commissioned for the transport of large quantities of water or foam concentrate, and these are known as water carriers or foam carriers. Control units (frequently now called incident command and control units) from which the officer in charge of a large incident can co-ordinate operations, foam tenders, hose layers and salvage tenders continue to play their part in the fight against fire. It is now commonplace for many of these special appliances to be provided in body-only form so that one vehicle, known as a prime mover, can transport the required body to an incident, leave it there and remain available for further deployment. Some brigades achieve the same result by the use of trailers.

By now we can see that fire engines are not all the same and some of them are not even designed to deal with fires at all. There are many variations on the theme to suit the particular needs of both public and works brigades, with appliances ranging from perhaps the 2 tonnes of a Land Rover to the 50 tonnes of an eight-wheel drive airport crash tender. This book can describe and show no more than a representative selection. For the reader who wants to pursue a general interest in fire engines and the fire service I recommend membership of the Fire Brigade Society. Anyone interested in the restoration and preservation of old, and not so old, appliances could do no better than join the Fire Service Preservation Group. My thanks are due to members of both organisations, and to the fire brigades and others who have helped and co-operated in the preparation of this book.

The convention employed when describing fire appliances is for the vehicle chassis to be mentioned first, followed by the bodybuilder/supplier and then the manufacturer of any special equipment or fire engineering. Imperial or metric units of measurement have been quoted generally according to whichever prevailed for construction of the appliance described. The only abbreviations used which some readers may not be familiar with are gpm — gallons per minute, and lpm — litres per minute.

Roger C. Mardon,
Canterbury, Kent.
December 1998

This page: This horse-drawn steamer was built in 1883 by Shand Mason and supplied to Bromley Urban District Council in 1887 at a cost of £430. It is a double-cylinder vertical steam pump producing 260gpm, the stroke of the piston in one of the two vertically mounted cylinders alternating with the other so that the water is always under pressure from one of the pistons. In 1919 the engine was sold for use at a Horton Kirby paper mill where it was apparently used to water the sports ground and not for fire-fighting. It has recently been fully restored and is now maintained as part of the Kent Fire Brigade museum.

Right: In 1929 Hove Fire Brigade purchased this Dennis Braidwood pump escape powered by a 45hp four-cylinder petrol engine. The pump has an output of 400/600gpm through four deliveries and a 100gal water tank and hose-reel are fitted. This was one of the first fire engines to be fitted with pneumatic tyres from new. It served the war years at Hove and after denationalisation of the fire service in 1948 became part of the East Sussex Fire Brigade fleet, serving at Newhaven until 1952. It was acquired by the Baker Perkins works fire brigade in Peterborough where it was on the run as a pump until 1971.

BOROUGH OF HOVE
PN-4119
THE KNOB
VINCENT
MAX SPEED 20M.P.H.
UW 4. 2. 0.

HAMPSHIRE
FIRE SERVICE
BOR 316

Left: This Dennis Light Six pump with Braidwood body was new to Aldershot Fire Brigade in 1936 and served there until 1952, having been absorbed by the National Fire Service in 1941 and taken into the Hampshire Fire Service fleet on denationalisation in 1948. It also provided cover in Farnborough for a short time and later in Hythe where it remained until 1961.

Below: A Leyland Cub FK7 pump escape with limousine body in the distinctive livery of Newcastle City Fire Brigade. The first Newcastle machine to be supplied with this livery was in 1933. Following the war and the return of the fire service to local authority control in 1948, Newcastle and Gateshead operated a joint fire brigade and continued to order appliances finished in maroon and red until 1968. Note the absence of doors to the crew compartment of this vehicle which was one of three identical appliances delivered in 1937.

In 1937 London Fire Brigade took delivery of two German-made Metz turntable ladders mounted on Leyland TLM chassis. Each was fitted with a Rees Rototurbo 500gpm pump enabling the monitor at the head of the ladder to be supplied with water without relying on another appliance. The Metz ladder extends to 101ft and has a full range of movements driven from a power take-off on the road engine gearbox and controlled by 10 clutches. The vehicle pictured was stationed at Soho for the whole of its working life until 1963.

The Bedford M commercial vehicle chassis was introduced in 1939 and, with the exception of the war years, remained in production until 1953. This appliance is one of a pair of Bedfords purchased in 1939, before manufacture was suspended, by Boston Rural District Council Fire Brigade to tow a Scammell trailer pump. It has a 200gal water tank but no built-in pump. After the war it became part of the Holland (Lincolnshire) Fire Brigade fleet and, following a brief spell as a hose layer in the 1950s, was taken out of service in 1956.

Above: Leyland FT4A pump escape new for Stafford Borough Fire Brigade in 1939 and named 'Eveline' by the mayor of the borough at the time, Eveline Joynes. This vehicle is powered by a six-cylinder 8,847cc petrol engine producing 49.5hp. The appliance has a Gwynne 1,000gpm pump mounted amidships, a 100gal water tank and carries a 50ft Ajax escape. After the war, during which it saw service in the blitz at Coventry, this vehicle passed to Staffordshire County Fire Brigade and remained in service until 1965.

Right: In comparison with the enclosed appliance (pictured on the cover) supplied to Bournemouth a year earlier, in 1940 Stourbridge Fire Brigade in Worcestershire chose open bodywork with transverse seating for its Leyland FK9 pump escape. It is equipped with a 500gpm Leyland Gwynne pump and 50gal water tank and carries a 60ft Bayley escape.

Winsford Fire
Station
LEYLAND
BOROUGH OF STOURBRIDGE
FIRE BRIGADE
DNP 721

The Dennis Light Four was introduced in 1936 and remained in production until the end of the war. The appliance pictured was new to the Massey Harris works fire brigade, Coventry in 1939 and exemplifies the 'New World' body style with the crew sitting inside an open body on inward facing seats. Other examples of the Light Four were built with an enclosed cab for the driver and officer in charge and open Braidwood bodywork behind for the crew. The pump is a rear-mounted Dennis No 2 producing 500/600gpm. This vehicle is now preserved by retained fire-fighters at Kenilworth.

In the early stages of World War 2, about 1,000 pumps were built on Bedford chassis and others on Morris Commercial chassis. These makes were soon required to fulfil orders for military vehicles and fire pumps were constructed on Austin K4 and Fordson 7V chassis. Most were capable of producing 700gpm through four delivery outlets and were designated heavy units. Some produced 1,100gpm through six deliveries and were designated extra heavy units. This Bedford heavy unit was built in 1940.

Left: The auxiliary towing vehicle was designed in the light of experience gained in the early stages of the war from the use of taxis and other light vehicles to tow trailer pumps. A 2-ton chassis with a steel body providing seating accommodation for the crew and internal locker space for equipment was adopted. The roof was specially strengthened to afford protection from shrapnel and other flying debris. This Austin K2, built in 1941 and seen in the wartime grey livery of the National Fire Service, is towing a Coventry Climax light trailer pump.

Left: Mobile dam unit built in 1943 for the National Fire Service on a Dodge 82A chassis which passed to Buckinghamshire Fire Brigade on denationalisation in 1948. These appliances, with a pump and 500 or more gallons of water on board, often comprised the first attendance at fires where water was unlikely to be available and their early deployment could prevent a fire getting out of control. They were also used to patrol areas where the fall of incendiary bombs had been reported and, because mains water supply was frequently disrupted during air raids, for damping down where the mains had failed.

This 1943 Austin K4/Merryweather 60ft hand-operated turntable ladder is one of 50 supplied to the National Fire Service in 1943/4. Generally speaking, this ladder was within its safe working limits at any angle of elevation and was designed to be used by crews who were not necessarily qualified as turntable ladder operators. Although not delivered with a pump, many of these appliances were later fitted with the American front-mounted Barton 300gpm pump. This particular example went to West Sussex Fire Brigade after the war and served at Chichester and Worthing, finally being withdrawn from service in 1970.

Oil fires are extinguished by the application of a layer of foam to exclude oxygen, without which fire cannot be sustained. Special risks such as oil refineries need to have ample quantities of foam available and this 1948 AEC Mammoth Major was put into service as a foam carrier, with 2,000gal of foam compound, at Esso Fawley Refinery in 1951. Also on the run with the works brigade there in 1977, when this picture was taken, were two 1,000gpm multi-purpose (foam/water/dry powder) appliances, a tailboard loader carrying six mobile monitors and a forward control unit.

The Dennis F.7, with its 13ft 6in wheelbase, was introduced in 1949 and discontinued in 1951. Early models are distinguished by their plain rectangular radiator but later versions were built with a chrome grille characteristic of the shorter F.12. Otherwise the F.7 can be recognised by the full-width locker between the pump bay and the rear wheel arch, which is narrower on the F.12. The pump escape pictured was supplied to the City of Chester Fire Service in 1949. It carries a Morris Ajax 50ft wheeled escape, 35ft and 17ft Ajax extension ladders and is fitted with a midships-mounted Dennis 1,000gpm pump and 100gal water tank.

Unusual with its inboard twin hose-reels and ladder nest, this 1951 type A water tender is one of six similar appliances built for Nottinghamshire Fire Service on an Austin K4 chassis converted to forward control. Originally with a light portable pump mounted at the rear and designed to tow a trailer pump, this one was later fitted with a front-mounted pump. It remained with the brigade until 1966 and in 1981 was in service with Billy Smart's Circus at its base and winter quarters at Winkfield, near Windsor.

The Dennis F.8, introduced in 1950, was a popular appliance with both city and rural brigades. Only 6ft 6in wide and 9ft 6in high complete with ladder, it was ideally suited for work in the confines of city streets or in country lanes. This unusual 1953 example is one of five built for Leeds City Fire Brigade with access to the crew accommodation from the back of the vehicle — note the absence of rear doors. It is fitted with a 150gal water tank and 500gpm pump and remained on the run until the early 1970s.

Left: After nearly seven years of nationalisation, the fire service reverted to local authority control in 1948. As part of the Government's response to the threat of nuclear war, the Civil Defence and Auxiliary Fire Service (AFS) were reconstituted in 1949 under the Home Office, with the intention that both would be made up of civilian volunteers. Emergency pumps were built in the 1950s to replace the wartime vehicles then available to the AFS. Despite having no military pedigree, they were painted green and became affectionately known as 'Green Goddesses'. This 1953 model typifies earlier versions built with a 400gal water tank on the Bedford SH rear-wheel-drive chassis. It originally served at West Hartlepool but passed to the Central Electricity Generating Board in Hartlepool after the disbanding of the AFS in 1968.

Below left: Later models were built on the Bedford RL chassis. A Sigmund 900gpm pump is rear-mounted and, on this four-wheel-drive appliance, a 300gal water tank and two 180ft hose-reels are fitted. A Coventry Climax light portable pump, 1,600ft of rubber-lined delivery hose and a 35ft aluminium extension ladder were carried as standard on these appliances.

Some AFS appliances were to be deployed in mobile fire columns of 144 vehicles intended to reinforce hard-pressed local services anywhere in the country. The Column Commander would have use of a mobile control unit designed to operate, with a crew of six, as communications centre, operations room and information exchange. The main body was fitted as an operations room with provision for maps and mobilising boards. Separate compartments were provided for two VHF radio sets and a field telephone switchboard with 10 lines. Shown is such a control unit on a 1957 Bedford RL chassis.

The Karrier Gamecock chassis, with its 9ft 7in wheelbase, provided a compact and manoeuvrable fire engine. Carmichael & Sons of Worcester built this 1956 water tender with a 500gpm rear-mounted pump and locker space for 2,000ft of 2½in delivery hose. A 35ft extension ladder and a short ladder are recessed in the roof and this feature, together with the low mounting of the bells, gives a height of only 8ft 7in. The appliance shown served with Kent Fire Brigade at Tenterden and is pictured at a display in Herne Bay while forming part of the brigade's museum.

Above: This Magirus 100ft turntable ladder is one of eight supplied to Kent Fire Brigade between 1957 and 1961 on Bedford S and Commer C7 chassis. The later models were hydraulically powered but this 1957 example supplied by John Morris & Sons Ltd of Salford on the Bedford chassis was mechanically operated, elevation being by means of a single screw drive. It spent its working life at Medway fire station until it was replaced by a hydraulic platform in 1980.

Right: This Dennis F.21/Metz DL37 mechanically operated turntable ladder was supplied to Rochdale Fire Brigade in 1957. It is one of only two such ladders supplied in the United Kingdom capable of extending to 125ft, the other having gone to Belfast three years earlier. The ladder is elevated, that is raised from the headrest, by twin chains affixed to the semi-circular swinging frame on which the ladder is mounted. This model is powered by a Rolls-Royce B.80 engine and is fitted with a Dennis 500gpm pump. It was taken into the Greater Manchester Fire Service in 1974 and remained in service until 1976.

Below: The Royal Air Force Fire Service dual purpose fire/crash tender Mark 1 (DP1), built by Foamite on a Thornycroft 4x4 chassis, was introduced in 1956 with the basic objective of supplying water to foam tenders. With equal efficiency it could be used as a secondary foam tender or as a self-contained domestic appliance for dealing with fires at RAF ground bases. It was powered by a Rolls-Royce B.80 engine of 140bhp and equipped with 700gal water and 35gal foam tanks. A Coventry Climax 350/500gpm pump was fitted together with a round-the-pump proportioner for foam generation.

Right: The Ministry of Transport & Civil Aviation Aerodrome Fire Service was responsible for providing fire and rescue services at Ministry-controlled airports in Great Britain and Northern Ireland until 1966, when responsibility transferred to the British Airports Authority. A typical appliance of the late 1950s and 1960s is this Thornycroft Nubian foam tender by Sun Engineering (Richmond) put into service in 1957 at Gatwick. It carried 800gal of water and 100gal of foam and was able to produce 7,000gpm of foam. Rescue tools and lighting equipment were also carried. The appliance was pictured after acquisition by Duxford Aviation Society, Cambridgeshire.

DUXFORD AIRFIELD
3
SXT 107
THORNYCROFT
103

DENNIS
MIDDLESEX
39 MMX
P130
67

*Left:*The Dennis F.12 was in production between 1950 and 1959, during which period 336 were built at the Dennis factory in Guildford, Surrey. This one was supplied to Middlesex Fire Brigade in 1958 and served as a pump escape at Southall until 1970, under the auspices of the London Fire Brigade from 1965, after which it was one of 50 pumps required at a Tottenham timber yard fire. On a 12ft 6in wheelbase, it is powered by a Rolls-Royce B.80 eight-cylinder 5.6-litre petrol engine and can accelerate from 0-60mph in 45sec. A Dennis 1,000gpm pump is mounted at the rear and a Bayley 50ft escape, two hook ladders and a short extension ladder are carried on the roof.

Below: This 1958 Ford Thames is an ex-Civil Defence vehicle acquired by East Sussex Fire Brigade in 1969 for use as a control unit. It bears the distinctive red and white chequer markings of such an appliance and a public address loudspeaker is mounted on the front bumper. It is seen here at Lewes fire station where it remained on the run until replaced by a new and much larger Bedford YRQ/Anglo Coachbuilders vehicle in 1978.

Left: From the AEC/Merryweather Marquis range, this foam tender for Kent Oil Refinery Fire Service was one of two supplied in 1959. Able to deliver water, water fog or foam, the appliances could be connected to the refinery's permanent pipe installation or used for independent attack. In addition to 500gal of foam compound, these appliances carried 200gal of water and were equipped with a 180ft hose-reel worked from the main pump, a 30ft ladder, breathing apparatus, resuscitator, asbestos blanket and gloves, rescue line, fog nozzles, ground monitor and a variety of hand extinguishers.

Below left: BP Refinery (Kent) Ltd also operated this 1960 emergency tender to protect a refinery complex of some three square miles. The cab provided accommodation for the driver and officer-in-charge, with seating for four crew members in the rear compartment which also provided a work bench and carried emergency gear. A central compartment provided a further work bench with vice and storage for two 11in searchlights powered by an 18-cell alkaline battery. Two water tanks holding a total of 150gal supplied the 30gpm first-aid pump and hose-reel. A portable pump was housed in an outside locker and a 25ft light alloy ladder was carried on the roof.

In 1960 Surrey Fire Brigade designed and built this breakdown lorry in its own workshops. The chassis was a Thornycroft Nubian with six-wheel-drive powered by a Rolls-Royce B.81 engine. The power winch was capable of hauling 20 tons from the front, rear or either side of the vehicle and the turntable crane at the rear could lift 5 tons. The appliance was designed for heavy rescue work as well as the recovery of brigade vehicles and carried an axle ambulance, 5-ton snatch block and two 15-ton lift jacks.

FIRE
ESSEX COUNTY FIRE BRIGADE
SOUTHEND
XHJ 219

Left: This pump escape was delivered to Southend-on-Sea Fire Brigade in 1960 and is seen here shortly after being absorbed into Essex County Fire Brigade in 1974. It is one of the Marquis range built by Merryweather on an AEC chassis and equipped with the company's own midships-mounted pump delivering 750gpm and all-steel 50ft wheeled escape weighing 10½cwt.

Above: This long-wheelbase Land Rover emergency tender was built by Carmichael of Worcester in 1961 to the requirements of Bath City Fire Brigade for road accident rescue. A front-mounted winch and a range of jacking equipment was provided along with oxy-acetylene cutting equipment. A 110V 5kW generator powered electrical equipment including a Black & Decker 9in angle grinder and a 14in gauge shear. Three Francis flood projector lamps and tripods were carried, one mounted on the roof for immediate use.

This 1961 Commer/Jennings water tender of Norfolk Fire Service is equipped with a Hayward Tyler 500gpm pump fitted with a water seal primer. When lifting water from an open source a pump must first be primed to exhaust any air in the suction hose and pump casing, allowing water to be forced into the pump by atmospheric pressure. The water seal primer achieves this by pumping out water always kept in the base of the pump which in turn entrains and draws out any air. It was not extensively used in fire pumps partly because the need to keep water in the pump created frost problems in winter.

To meet the requirement for high-speed appliances capable of negotiating all types of terrain in all weather conditions, the Alvis Salamander six-wheel-drive crash tender was developed from an armoured fighting vehicle chassis/hull. Built by Pyrene with the company's own foam-producing equipment installed, this tender had chlorobromomethane (CBM) equipment by General Fire Appliance Co and was powered by a Rolls-Royce B.81 engine mounted at the rear. It was taken into service by the Royal Air Force and designated the Mark 6 crash tender. The appliance shown is a Mark 6C with dual output 2,500/5,000gpm monitor commissioned in 1962.

Below: The Bedford TK was a popular chassis for fire engines from its introduction in 1960 until it was discontinued 24 years later. When this Bedford TK/HCB-Angus pump escape was new in 1964 the county boroughs of Brighton, Eastbourne and Hastings operated their own brigades and East Sussex Fire Brigade had only one whole-time station, at Hove. Lewes, where this picture was taken in 1974, was one of the county brigade's eight day-manning stations, 16 others being crewed by part-time retained personnel.

Right: ERF Ltd of Sandbach, Cheshire exhibited its first fire appliances at the Commercial Motor Show in 1966. Its 84RS chassis was designed to carry a hydraulic platform and the 84PF was designed for a pumping appliance. This 1968 ERF/HCB-Angus water tender escape, fitted with the 400gal water tank required for a water tender and carrying a 50ft Merryweather all-steel wheeled escape, served with Nottinghamshire Fire Brigade at Newark. In addition to the 500gpm main pump, a Hathaway light portable pump was carried. Connections were provided for the operation of air-powered cutting and rescue tools using the air reservoir of the braking system.

FIRE
NOTTINGHAMSHIRE
NEWARK
ERF
SAL 999F

BAYLEY
HAMPSHIRE FIRE BRIGADE
BEAULIEU
BEDFORD
LHO 512F

Left: From its inception in 1948 Hampshire Fire Service built fire appliances in its own workshops, initially converting and improving wartime machines but later designing and constructing entire appliances on bought-in chassis. This achieved a significant saving when compared with the cost of buying appliances from commercial suppliers. In 1968 this water tender, one of 37 built on the Bedford TKEL chassis between 1962 and 1972, was completed and allocated to Beaulieu. The successful design was, of course, unique to Hampshire.

Below: This AEC/Merryweather Marquis pump ladder was new to the City of Sheffield Fire Brigade in 1970. It is seen displayed at the exhibition of appliances and fire-fighting equipment which accompanied the annual conferences of the Chief Fire Officers Association and the Institution of Fire Engineers that year in Folkestone. The appliance carried a 45ft ladder from new and was provided with a 1,000gpm rear-mounted pump and 400gal water tank.

Right: The 'Scoosher' was originally developed by Glasgow Fire Service in 1968 as a lightweight appliance capable of remotely directing a water jet into areas of difficult access from the monitor at the head of its articulated hydraulic boom. The 1970 vehicle shown is the heavier Mark 2 version bodied by J. C. Bennett & Co (Coachbuilders) Ltd of Glasgow on a Dennis F.46A chassis with a 1,000gpm pump and 300gal tank. A 46ft 4in Lacon ladder permanently mounted alongside the lower boom of the Simon 45ft unit was extended by a power winch and protected by safety devices.

Left: Yellow fire engines were first seen on the streets of Britain in 1966 after Coventry Fire Brigade introduced them following experiments at the Lanchester College of Technology. The objective was to find a colour which presented a unique and distinctive appearance in daylight yet retained its colour value under street lighting. Yellow, with a wavelength between 5,300 and 5,500 angstroms for the technically minded, was found to meet the requirement best and West Sussex Fire Brigade was one of those to adopt the new colour with this Rolls-Royce-powered 1971 Dennis F.44 water tender ladder stationed at Crawley.

Below left: Another colour adopted for its conspicuity was white, as demonstrated by this water tender of Bedfordshire Fire Service built in 1972 by Merryweather on a Ford D series chassis. This appliance was stationed at Kempston but when the photograph was taken in 1980 newer appliances there, and elsewhere in the county, were finished in red.

This was one of two similar rescue tenders operated by Somerset Fire Brigade to fulfil the need for a small, fast equipment carrier to supplement attendance at road traffic accidents. It was a Ford Transit 190 van powered by a 3-litre V6 petrol engine with engineering for brigade use by HCB-Angus. Air tools and lifting bags were powered from a Hydrovane compressor driven through a power take-off. Additional equipment included Epco hydraulic tools, trolley jacks and blocks. This appliance remained on the run at Taunton from 1974 until replaced by a Reynolds Boughton RB44/Saxon Sanbec in 1984.

Below: The Redwing range of Land Rover appliances produced by Carmichael offered a choice of pumps up to 500gpm which could be rear-mounted or midships-mounted. The FT.6 came with an all-aluminium body, fibreglass roof and drop-down locker doors. Distinguished by its forward-control configuration, the appliance provided accommodation for a crew of four with a four-door cab for ease of access. This 1975 example has a 100gal water tank and was originally fitted with a 25gpm gear pump to supply the hose-reel. It was later equipped with a 250gpm Angus portable pump and was still in service with Cheshire Fire Brigade at Nantwich 23 years after it was built.

Right: In the late 1960s the City of Plymouth Fire Brigade developed an appliance that encompassed the roles of pump, rescue, salvage and foam tender. Plymouth and Exeter city brigades amalgamated with Devon County Fire Service in 1973 and the multi-purpose appliance soon became the new Devon Fire Brigade standard, although that cumbersome designation was dropped in favour of water tender ladder in 1981. This Dennis D appliance was supplied in 1976 and served at Ottery St Mary.

DENNIS
OTTERY ST. MARY
DEVON
FIRE BRIGADE
411
LTT 208P

The Royal Air Force TACR — truck airfield crash rescue — was designed primarily to reach the scene of a crash as quickly as possible to effect a rescue. The Mark 1 was built on a Land Rover chassis but the TACR 2, as depicted here, was built on a 6x4 Range Rover conversion with trailing rear axle to carry 900 litres of aqueous film-forming foam (AFFF) for fire-fighting. Some examples were built by HCB-Angus but most, like this one put into service in 1977, were constructed by Gloster Saro. From 1985 the TACR 2A version came from the Carmichael stable.

The business of G&T Fire Control Ltd was started in 1972 for the overhaul and modification of turntable ladders. Recognising that the life of a ladder and its mechanism could be twice that of the chassis on which it was mounted, five years later the company started to undertake re-chassis conversion work. This Merryweather ladder set was re-chassised from a Bedford TKG on to an ERF 84PS for Bedfordshire Fire Service in 1978, making it one of only a handful of turntable ladders to be mounted on an ERF chassis.

This Dennis F.125/Dennis/Simon SS263 hydraulic platform was put into service by Cheshire Fire Brigade in 1980. The Simon Snorkel SS263 three-boom unit provides a working height of 27.8m (91ft), sufficient to reach ninth-storey windows. The maximum safe load of the cage is 365kg (800lb) and a maximum outreach of 14.5m (47.6ft) is possible. Automatic limit stops operate on all controls making it impossible to exceed safe working limits. The appliance is seen at Macclesfield standing in for its sister machine normally on the run there.

The EPL Firecracker 235 hydraulic platform was first seen at the Firetech '79 exhibition. It was of compact design with a telescopic upper boom and achieved an operational height of 23.5m. Kent Fire Brigade took delivery of three Firecrackers mounted on Dennis F.125 chassis in 1981 but they remained in service at Dover, Maidstone and Tunbridge Wells for a short time only. The brigade had ordered chassis suitable for either hydraulic platforms or turntable ladders and in the same year three Magirus ladders, delivered by Carmichael, were also added to the fleet.

HAMPSHIRE FIRE & RESCUE SERVICE
DENNIS
2
FIRE
BTP 490W

Left: Hampshire's tradition of building many of its own appliances continued after the county fire service merged with the Southampton and Portsmouth brigades, to form the Hampshire Fire Brigade, as a result of local government reorganisation in 1974. This 1981 water tender ladder represents the later period of appliance construction in Hampshire's own workshops. It is based on a Dennis RS chassis and was originally allocated to Copnor. No pumping appliances were built by the brigade after 1989 but the Hampshire Fire and Rescue Service of today still fits out and completes new machines delivered by commercial suppliers.

Above: Northumberland Fire and Rescue Service had this 1981 Dodge G1313/Carmichael water tender ladder on the run at Morpeth in 1996. Morpeth is the administrative centre of Northumberland and the headquarters of the county fire and rescue service is there. The fire station, crewed by whole-time and retained personnel, supported a water tender ladder with rescue capabilities offered by Clan Lukas hydraulic cutting equipment as well as this appliance which carried either Combi Tool or Epco hydraulic rescue gear.

LEICESTERSHIRE FIRE & RESCUE SERVICE
EMERGENCY TENDER
Angloco
DODGE
LOUGHBOROUGH
VUT 311X
204

Left: In 1981 Leicestershire Fire Service commissioned two emergency tender/control units on Dodge Commando G13 chassis with bodywork by Angloco. They were each equipped with a 10kW 110V generator, 5.3-tonne winch and 4.5m extending mast light. Equipment included cutting and spreading gear, air bags capable of lifting up to 40 tonnes, gas-tight suits, radiation equipment and eight breathing apparatus sets. A control unit area at the rear accommodated two radio operators for the brigade main scheme, with links to five neighbouring brigades, and the fireground radio scheme. This appliance was allocated to Loughborough.

Below: This 30m Magirus turntable ladder was built on a Dennis DF133 chassis with Perkins V8-640 215bhp engine and Allison MT.643 automatic gearbox. It was supplied by Carmichael to Durham Fire Brigade in 1983 and is seen here as a reserve machine at Newton Aycliffe, having previously served at Darlington until replaced by a Volvo FL10/Angloco/Bronto Skylift F32HDT aerial ladder platform in 1996.

Below: Saxon SVB, now Saxon Sanbec, was a company established by former employees of the ERF subsidiary, Cheshire Fire Engineering, which closed its factory in 1982. Saxon soon won orders to supply Bedford appliances to Somerset Fire Brigade and this water tender ladder was completed on the TKG chassis in 1983. It has an Allison AT545 automatic gearbox and is fitted with a rear-mounted Godiva UMPX 2,270lpm pump and 1,800-litre water tank. After serving at Bridgwater until 1987, it was moved to Chard and later into the reserve fleet.

Right: A London Fire Brigade medium demountable chassis based on a 1984 Bedford CF 350L with Ray Smith running frame and Wadham Stringer fire investigation unit body. Five fire investigation teams are established in London and are available to determine the cause of any fire. A team is automatically sent on to any fire resulting in a fatality, fires where persons are reported unaccounted for and those where pumps are made up to four or more. The officer in charge of any incident can request the attendance of a fire investigation team and will normally do so for fires of doubtful origin.

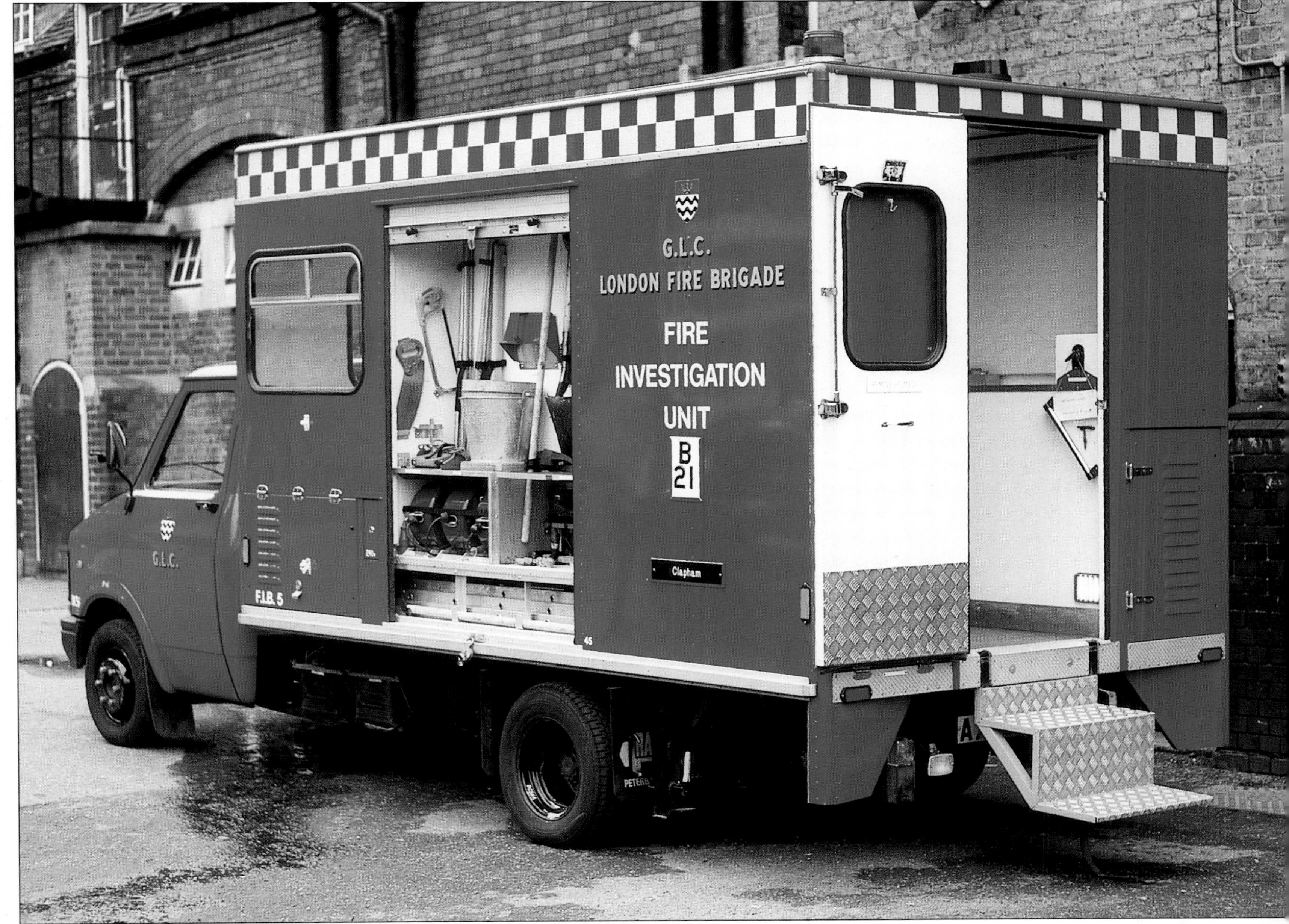
G.L.C.
LONDON FIRE BRIGADE
FIRE
INVESTIGATION
UNIT
B
21
Clapham
G.L.C.
F.I.B. 5

GLOUCESTERSHIRE FIRE & RESCUE SERVICE
STROUD
G 13
CARMICHAEL
B183 SDF

Left: This 1984 water tender ladder of Gloucestershire Fire & Rescue Service was built on a Dodge G13C chassis and powered by a V8 turbocharged engine to cope better with the hills in Stroud where it is seen. Bodywork was by Carmichael but in 1995 the body was removed and the vehicle was converted to a prime mover by Watts of Lidney using the lifting gear from one of the service's Bedford TKGs. The appliance in its new form is now used primarily to transport the brigade's demountable incident support unit based at Cirencester.

Above: This Bedford TL1630/Magirus DL30 turntable ladder extending to 30m was supplied in 1985 by Carmichael to Northamptonshire Fire Brigade. It is pictured six years later outside the fire station at Kettering, where it replaced a 1965 Bedford KG/Haydon/Magirus. Carmichael of Worcester held the franchise for Magirus ladders, in succession to David Haydon of Birmingham, from 1966 until 1992.

FIRE & RESCUE
CAMBERLEY
D423 WPG
POLYMA
Fire Protection
AUTOMATIC

Left: Surrey Fire Brigade took delivery of 21 Mercedes-Benz 1222F/Polyma water tender ladders in 1987. Each was equipped with a front-mounted 4-tonne electric winch and a hydraulically-operated 7m floodlight mast with four 1,000W halogen lights powered by a 110V generator. The lighting head, which could be operated at ground level from the appliance or by remote control, was fully directional through 360° and could be angled through 180° for maximum illumination of an incident. This appliance is pictured at Camberley in 1988.

Below: The primary role of this Avon Fire Brigade 1986 Renault-Dodge G13/Saxon chemical incident unit is the decontamination of personnel. It carries 12 chemical protection suits and nine gas-tight suits for the use of fire-fighters involved in an incident and showers for washing off hazardous substances. The appliance is also equipped with fax and telephone for the communication of technical information about the materials involved so that an accurate assessment of the risks to health and safety, and the environment, can be made.

Leyland Motors Ltd started building fire engines in 1909 and remained prominent in the field until the 1940s. The company returned to the specialist scene in 1958 with the short-lived Leyland Firemaster chassis. Thirty years after that, Scottish fire engine builders Fulton & Wylie constructed this 1988 water tender ladder for Gloucestershire Fire and Rescue Service on a Leyland Freighter 16.17 chassis. It was one of the first Gloucestershire machines to carry a triple extension 10.5m ladder, on the nearside, in addition to the 13.5m ladder and is seen here on the run at Cheltenham soon after delivery.

The Mercedes-Benz 917AF chassis is more usually seen as a rescue tender or support vehicle. Whilst Dorset Fire Brigade is not the only brigade to have ordered this chassis for use as a compact or mini water tender, it was the first. The water tank capacity is 1,363 litres, as opposed to the 1,800 litres of a full-size water tender, and the pump capacity is 2,270lpm. The ladder is 10.5m in three sections rather than two in order to reduce the overall length. The appliance depicted here was built by HCB-Angus as a demonstrator in 1988 but acquired by the brigade in 1989 and initially stationed at Ferndown.

Below: Telehoist of Cheltenham built both fire and rescue vehicles on GMC four-wheel-drive chassis powered by 6.2-litre V8 diesel engines. Buckinghamshire Fire and Rescue Service was among those to put light pumping appliances into service and this 1988 model is seen on the run at Buckingham in 1996. A 2,270lpm rear-mounted pump and 900-litre water tank were installed and the all-steel crew cab would accommodate five fire-fighters. In contrast to the cab, the bodywork was constructed of lightweight glass-reinforced plastic.

Right: The Gloster Saro Javelin crash tender, built on a Reynolds Boughton Taurus 6x6 chassis, carried 10,000 litres of water, 1,200 litres of foam and a Perren 100kg BCF (bromochlorodifluoromethane) unit. This 1988 appliance at Stansted Airport is seen with a hydraulically-operated Sky-King ladder mounted on the rear which extends to a height of 10.5m, enabling BCF extinguishant to be discharged into high-mounted aircraft engines such as those on the TriStar and DC-10. A high-level telescopic lighting mast provides illumination at the scene of an incident.

B·A·A
Stansted
Airport Fire Service
3
GLOSTER
JAVELIN
E189 KDF
GLOSTER SARO
04910

Three of these six-wheel-drive Land Rover/HCB-Angus pumps were built for Dorset Fire Brigade in 1988 and went on the run at Lyme Regis, Sturminster Newton and Gillingham, where this one remained until 1997. The appliance was fitted with an Albany AP8 pump and a 680-litre water tank. A 10.5m triple extension ladder was carried on the roof. Since coming off the run as a pump, the Lyme Regis machine has been converted into an environmental support unit and it is likely the other two vehicles will be similarly converted.

In 1990 Cheshire Fire Brigade commissioned this articulated control unit/canteen van based on a Leyland Roadrunner tractor unit and semi-trailer by A. L. Gordon of Falkirk. It would be ordered on to fires of five pumps or more and was manned by a pump crew responding with it. The main part of the body provided a control and command facility but the rear of the trailer offered canteen facilities beneath the hinged tail flap. This Winsford-based appliance, one of two similar vehicles operated by the brigade, was converted into a fire safety unit in 1998. The other has been refurbished and remains on the run as a control unit at Northwich.

COMMAND
FIRE
FIRE
917
Western Command
NW
2
CSU 2
BOOKING IN
COMMAND SUPPORT UN
H862 XYF

Left: In 1991 London Fire Brigade commissioned six light rescue units on the Mercedes-Benz 917AF chassis with bodywork by Locomotors. In 1993 they were converted by the brigade's Ruislip workshops to command support units and based at each of the Area Command Headquarters. They carry nominal roll boards for keeping a checklist of brigade personnel in attendance at an incident and are on the predetermined first attendance for some locations, particularly London Underground stations with more than one line. Otherwise, they are automatically ordered on to four-pump make-ups and above. A larger command unit attends six-pump make-ups but these arrangements will change with the bringing into use of a new brigade command unit in 1999.

Above: This Scania G93ML-250/Reliance Mercury rescue pump was delivered to Buckinghamshire Fire and Rescue Service in November 1991 and first stationed at High Wycombe. It was pictured in 1996 while in service at Broughton, one of the fire stations serving Milton Keynes. The appliance carries a 13.5m ladder and accident rescue capability is provided by Clan Lukas hydraulic cutting and spreading equipment.

This operational support unit was built for Avon County Fire Brigade in 1991 with bodywork by W. H. Bence on a Renault G300-24D chassis. It carries eight palletised 1,000-litre bulk packs of foam which are loaded and unloaded by the Moffet Mounty M2003 fork-lift truck mounted on the back of the vehicle. Also available for use as required are pallets of salvage equipment, a fuel pallet with tank and jerry cans for refuelling appliances and a refreshment pallet to provide canteen facilities at protracted incidents. It was pictured at Avonmouth fire station.

This Buckinghamshire Fire and Rescue Service water carrier *(Right)* was built in 1992 on a Volvo FL6.17 chassis with a Dairy Crest tank and allocated to Princes Risborough. In 1995 the appliance was rebuilt by GB Fire as a rescue pump *(Below)* equipped with a Godiva 4,546lpm pump and a Snozzle P50 telescopic boom. The 15m boom has a piercing nozzle originally designed for use in aircraft fire-fighting and rescue but it is capable of penetrating a building. The rebuilt appliance is pictured on the run at Aylesbury.

Left: London Fire Brigade employs medium and heavy prime mover chassis for its demountable bodies. Bodies for the heavy chassis include damage control, hose laying, operational support, fire conference and breathing apparatus (BA) training units, as well as this 1994 Saxon Sanbec BA major incident unit seen at Edmonton on a 1993 Volvo FL6.14/Locomotors prime mover.

Below: Each body can be raised on four legs permitting the prime mover to be reversed into position beneath it and, once correctly lined up, the legs are retracted to allow the body to take its position on the chassis. Other demountable bodies are exemplified by the 1992 Reynolds Boughton bulk foam unit and 1987 Locomotors command unit shown here.

This Carmichael Cougar crash tender built on a Timoney 8x8 chassis was handed over to the British Airports Authority Fire Service at Stansted Airport in July 1993. It is powered by an 18-litre Detroit diesel producing 825bhp (613kW) and accelerates from 0-50mph (0-80kmph) within 32sec. It carries 12,500 litres of water and 1,500 litres of foam. The pump can produce 10,000lpm and the roof monitor will throw a jet 60m at the rate of 4,550lpm.

Royal Berkshire Fire and Rescue Service put this rescue support vehicle on the run at Newbury in 1993. It was built by Locomotors on a Mercedes-Benz 1124AF chassis with a Palfinger crane mounted at the rear. Equipped primarily to cope with road accidents, and stationed for convenient access to the M4 motorway, it carries a wide range of tools in slide-out trays including Lukas rescue equipment and air lifting bags. A scaffold tower is carried on the roof and this can be assembled to afford rescuers the high working position sometimes necessary to release persons trapped in the wreckage of an accident.

The Channel Tunnel poses special risks for fire-fighting and rescue, and purpose-designed vehicles have been built to operate in the service tunnel between the two running tunnels. The Service Tunnel Transport System (STTS) is based on 1993 Mercedes-powered units with a driving cab at each end and interchangeable body units for fire, ambulance, police and maintenance purposes. Four STTS fire appliances, with bodies by John Dennis Coachbuilders, are stationed at each of the English and French terminals, along with two ambulance vehicles and one police communications unit. Two First Line of Response (FLOR) appliances, each with a crew of four, carry breathing apparatus (BA), fire-fighting, rescue and first-aid equipment. Two Second Line of Response (SLOR) vehicles, with accommodation and BA sets for 12 crew, transport personnel and equipment from supporting Kent and Calais appliances, with one (STTS 3) operating as a forward control unit.

FLOR appliances can attend any incident throughout the full length of the tunnel but SLOR vehicles cannot cross the English/French border until requested after an incident has been made bi-national. Six pumps initially respond from Kent Fire Brigade to tunnel incidents on the English side with a seventh attending the Incident Command Centre on the terminal site. Crews and whatever equipment may be needed are transferred to the SLOR appliances for transport through the service tunnel to the incident and STTS 4 is available to shuttle personnel and equipment to and from the terminal Emergency Response Centre. Alternatively road appliances and their crews can be ferried to an incident through the running tunnel on a goods vehicle shuttle train.

WEST SUSSEX FIRE BRIGADE
DENNIS
Lancing
FIRE
RESCUE
EMERGENCY
P824 JPO
Rapier

Left: West Sussex Fire Brigade introduced this high visibility colour scheme for its appliances in 1995. This Dennis Rapier SFD122 water tender ladder, with bodywork by John Dennis Coachbuilders, is equipped with high pressure hose-reels and a 1,800-litre water tank. In addition to a range of fire-fighting gear, it carries hydraulic rescue equipment, chemical protection suits and a generator. It was new to the brigade in 1996 and allocated to Lancing fire station.

Below: In the 1960s and '70s many brigades commissioned appliances known as pump hydraulic platforms — pumps with a 50ft (15m) Simon hydraulic platform. Wiltshire Fire Brigade is now alone in running the modern and bigger equivalent and has two in service. This one at Trowbridge was built in 1995 on an ERF EC10 chassis by GB Fire and equipped with an Interlift 25m platform built and installed by Cambourne Engineering. The pump will deliver 2,270lpm and the water tank has a capacity of 1,000 litres.

SURREY FIRE & RESCUE SERVICE
ANGUS SACOL
S218
M81 KPA
999
PAINSHILL
Wap turbo

Left: Surrey Fire and Rescue Service brought this salvage tender into use in 1995. It is a Dennis SS241 with bodywork by John Dennis Coachbuilders and operates from Painshill fire station, Cobham. It carries a 10.5m ladder and is equipped with a 110V generator to power lighting equipment and tools. Both a light portable pump and a deep-lift submersible pump are carried along with a high expansion foam generator, the fan of which can be reversed for smoke extraction purposes. Additional equipment includes the salvage sheets, buckets and mops necessary for protecting property and limiting water damage.

Above: This water tender ladder is one of 18 delivered to Devon Fire and Rescue Service in 1995. Its MAN L2000 chassis and Saxon Sanbec bodywork together make the manoeuvrable appliance required for many of Devon's country lanes. A Godiva 2,700lpm multi-pressure pump and a 1,800-litre water tank are fitted and the machine carries a 13.5m ladder. It is based at Bere Alston retained fire station.

Although operated by Avon Fire Brigade, this unusual appliance is owned by Railtrack and designed for use in the Severn Tunnel. Rail guidance gear fitted to the rear and centre of the vehicle allows it to transfer from road to rail, enabling a fast response to incidents in the tunnel. 12.5 tonnes of equipment can be carried on pallets which are loaded and unloaded with a Moffet Mounty fork-lift truck carried on the rear. Personnel and equipment can be rapidly moved within the tunnel by two two-seater Alumi rail-karts and trailers. The vehicle, a 1995 Renault Maxter G300.26D with curtain-sided bodywork by A. G. Bracey of Bristol, is pictured at Avonmouth fire station.

Kent Fire Brigade was the first in the country to take delivery of the distinctive low profile Iveco Magirus/GB Fire/Magirus DLK 23-12 turntable ladder. This is a 30m ladder of compact design with four stabilising jacks which can be extended individually to different positions, wholly within the width of the appliance if necessary. The two-man rescue cage moves automatically from the driving position, where it does not interfere with the driver's view of the road, to the working position when the jacks are operated. The first appliance was put on the run at Maidstone in 1996 followed by one at Canterbury and this one at Folkestone in 1997.

DENNIS
P246 VMJ
Sabre
Hertfordshire
COUNTY COUNCIL
Fire & Rescue Service
WATFORD

Left: Unlike the Rapier which uses Dennis Specialist Vehicles' 'space frame', a welded structure chassis, the Dennis Sabre evolved from the RS and SS chassis design and was launched in 1995. The first one went into service with Wiltshire Fire Brigade but this 1997 Sabre water tender ladder is seen on the run with Hertfordshire Fire and Rescue Service at Watford fire station. It was one of 15 Sabres then in the county with bodywork by John Dennis Coachbuilders.

Above: This Volvo FL10/Angloco/Bronto Skylift F32 HDT aerial ladder platform was new to Worthing fire station of West Sussex Fire Brigade in 1998. It has a working height of 32m and this picture clearly shows the rescue ladder alongside the telescopic booms. In addition to the fire-fighting and rescue roles of this appliance, the elevated platform offers the officer in charge of an incident a commanding view of the fireground. Operating with a crew of two, it has a computerised automatic jacking and load management system and is equipped with a remote control water monitor, stretcher and stretcher-carrier.